Answers to
BASIC SKILLS IN
MATHEMATICS/Book 2

R W Fox

Deputy Headmaster, Fort Luton
Secondary School for Boys, Chatham, Kent

Edward Arnold

© R W Fox 1975

First published 1975
by Edward Arnold (Publishers) Ltd
41 Bedford Square
London WC1B 3DQ

Reprinted 1978, 1979

ISBN: 0 7131 1980 2

Printed in Great Britain by
Unwin Brothers Ltd, Woking

P. 1   Table work

0	0	0	0	0	0	0	0	0	0	0	0
1	2	3	4	5	6	7	8	9	10	11	12
2	4	6	8	10	12	14	16	18	20	22	24
3	6	9	12	15	18	21	24	27	30	33	36
4	8	12	16	20	24	28	32	36	40	44	48
5	10	15	20	25	30	35	40	45	50	55	60
6	12	18	24	30	36	42	48	54	60	66	72
7	14	21	28	35	42	49	56	63	70	77	84
8	16	24	32	40	48	56	64	72	80	88	96
9	18	27	36	45	54	63	72	81	90	99	108
10	20	30	40	50	60	70	80	90	100	110	120
11	22	33	44	55	66	77	88	99	110	121	132
12	24	36	48	60	72	84	96	108	120	132	144

P. 2   Easy revision tests

	Test 1	Test 2	Test 3	Test 4	Test 5
1.	11	11	13	13	12
2.	7	8	9	5	9
3.	40	63	32	35	54
4.	4	5	6	3	7
5.	3a	4a	6a	8a	2x
6.	5b	3b	5b	b	y
7.	300	80	500	240	400
8.	50	10	10	1	5
9.	£2.40	£3.90	£3	£6	£1.20
10.	60p	40p	30p	30p	60p

	Test 6	Test 7	Test 8	Test 9	Test 10
1.	12	12	14	13	15
2.	3	8	7	6	7
3.	45	28	24	27	56
4.	9	5	5	6	7
5.	4x	4x	5x	5a	5a
6.	2y	0	2y	3b	2b
7.	200	500	200	200	300
8.	4	10	1	6	40
9.	£1	£1.80	£3	£4.40	£7
10.	30p	20p	50p	£1.20	60p

P. 3	Test 11	Test 12	Test 13	Test 14	Test 15
1.	15	16	21	21	23
2.	9	6	7	8	9
3.	60	48	48	72	42
4.	7	9	9	7	12
5.	2a	7a	a + b	7a	7a
6.	b	1	0	3x	2x − y
7.	4000	2000	3000	5000	1000
8.	20	1	2	10	600
9.	£5	£9	£4.60	£11.25	£8.10
10.	70p	30p	64p	32p	24p

	Test 16	Test 17	Test 18	Test 19	Test 20
1.	23	32	32	42	44
2.	15	17	15	17	18
3.	55	64	108	121	84
4.	8	12	11	9	6
5.	x + y	6x	a + 2b	7x	4a + b
6.	4x	4a	3a −2b	0	3a
7.	5000	600	2000	10 000	10 000
8.	10	10	2	20	200
9.	£15.05	£3.60	£4.76	£3.24	£3.78
10.	54p	93p	£2.41	£1.99	£1.37

P. 4   EX. 1

1.	True	2.	False	3.	True	4.	False	5.	False
6.	True	7.	True	8.	True	9.	False	10.	False
11.	True	12.	False	13.	False	14.	True	15.	True
16.	False	17.	False	18.	True	19.	False	20.	True

P. 4   EX. 2

1.	960	2.	511	3.	432	4.	1792	5.	1050
6.	2091	7.	1628	8.	1476	9.	2924	10.	1050
11.	1830	12.	2691	13.	14	14.	15	15.	18
16.	20	17.	40	18.	35	19.	41	20.	27
21.	33	22.	18	23.	39	24.	24		

P. 5   EX. 3

1.	23 R.3	2.	58 R.3	3.	62 R.2	4.	46 R.5
5.	48 R.0	6.	26 R.1	7.	23 R.6	8.	30 R.1
9.	36 R.6	10.	18 R.0	11.	12 R.7	12.	16 R.9
13.	54 R.0	14.	19 R.10	15.	16 R.6	16.	23 R.0
17.	61 R.5	18.	43 R.12	19.	86 R.0	20.	64 R.8
21.	58 R.20	22.	64 R.33	23.	105 R.41	24.	237 R.65

P. 7   EX. 4

1.	10 yr	2.	20	3.	100	4.	70 yr	5.	900
6.	248	7.	734	8.	184	9.	217	10.	204
11.	5	12.	14	13.	58	14.	170	15.	462
16.	648	17.	380	18.	9715	19.	23 yr	20.	103
21.	15	22.	26	23.	694	24.	10	25.	600 miles
26.	50 g	27.	23 R.45 g	28.	1332	29.	864	30.	252

P. 10   EX. 5

1.	9a	2.	6b	3.	15c	4.	18x	5.	12y
6.	18a	7.	2x	8.	3y	9.	3a	10.	12x
11.	5b	12.	c	13.	5x	14.	4y	15.	8b
16.	6b	17.	6a	18.	4x	19.	abc	20.	xyz
21.	2abc	22.	2ab	23.	2abc	24.	2abc	25.	2ab
26.	2ab	27.	2abc	28.	2abc	29.	4abc	30.	4abc
31.	2abc	32.	2abc	33.	2abc	34.	2abc	35.	2abc
36.	2abc	37.	2abc	38.	2abc	39.	4ab	40.	6ab
41.	6xy	42.	9xy	43.	24abc	44.	xyz	45.	60xyz
46.	abc	47.	2bc	48.	2ac	49.	2ab	50.	2c
51.	2b	52.	2a	53.	2	54.	1	55.	4abc
56.	0	57.	2abc	58.	2abc	59.	2b	60.	8abc

P. 11   EX. 6

1.	6	2.	8	3.	24	4.	0	5.	0	6.	12
7.	12	8.	0	9.	12	10.	36	11.	0	12.	8
13.	24	14.	96	15.	0	16.	30	17.	5	18.	6
19.	7	20.	1	21.	2	22.	2	23.	1	24.	2
25.	1	26.	2	27.	1	28.	3	29.	3	30.	0
31.	0	32.	0	33.	16	34.	12	35.	0	36.	24

P. 11 EX. 7

1. 6	**2.** 3	**3.** 1	**4.** 2	**5.** 1	**6.** 5
7. 1	**8.** 3	**9.** 2	**10.** 144	**11.** 9	**12.** 1
13. 14	**14.** 1	**15.** 26	**16.** 4	**17.** 6	**18.** 72
19. 9	**20.** 8	**21.** 30	**22.** 18	**23.** 2	**24.** 15

P. 13 EX. 8

1. 120	**2.** 180	**3.** 240	**4.** 300	**5.** 360	**6.** 420
7. 480	**8.** 540	**9.** 600	**10.** 1200	**11.** 1800	**12.** 2400
13. 3000	**14.** 3600	**15.** 3600	**16.** 3900	**17.** 4500	**18.** 2100
19. 2700	**20.** 8700	**21.** 180	**22.** 360	**23.** 540	**24.** 720
25. 1440	**26.** 320	**27.** 640	**28.** 870	**29.** 1245	**30.** 1435

P. 13 EX. 9

1. 20	**2.** 55	**3.** 30	**4.** 45	**5.** 25	**6.** 50
7. 10	**8.** 5	**9.** 120	**10.** 75	**11.** 15	**12.** 30
13. 15	**14.** 30	**15.** 45	**16.** 40	**17.** 30	**18.** 35
19. 35	**20.** 35	**21.** 40	**22.** 40	**23.** 45	**24.** 50
25. 45	**26.** 30	**27.** 55	**28.** 40	**29.** 40	**30.** 40

P. 14 EX. 10

1. 1 h	**2.** 1 h	**3.** 1 h	**4.** 2 h
5. 3 h	**6.** 3 h	**7.** 1 h 15 min	**8.** 1 h 15 min
9. 2 h 15 min	**10.** 1 h 45 min	**11.** 1 h 30 min	**12.** 1 h 20 min
13. 1 h 10 min	**14.** 1 h 30 min	**15.** 1 h 50 min	**16.** 1 h 40 min
17. 1 h 55 min	**18.** 1 h 35 min	**19.** 2 h 40 min	**20.** 2 h 50 min
21. 1 h 30 min	**22.** 2 h 10 min	**23.** 2 h 45 min	**24.** 2 h 40 min
25. 2 h 40 min	**26.** 2 h 18 min	**27.** 2 h 24 min	**28.** 2 h 38 min
29. 1 h 16 min	**30.** 2 h 25 min		

P. 15 EX. 11

1. 1.30 a.m.	**2.** 3.15 a.m.	**3.** 2.45 a.m.	**4.** 4.30 a.m.
5. 6.00 a.m.	**6.** 7.45 a.m.	**7.** 8.20 a.m.	**8.** 9.40 a.m.
9. 10.00 a.m.	**10.** 11.10 a.m.	**11.** 12 noon	**12.** 1.00 p.m.
13. 2.00 p.m.	**14.** 3.00 p.m.	**15.** 4.00 p.m.	**16.** 5.00 p.m.
17. 6.00 p.m.	**18.** 7.00 p.m.	**19.** 8.00 p.m.	**20.** 9.00 p.m.
21. 10.00 p.m.	**22.** 11.00 p.m.	**23.** 12 midnight	**24.** 1.15 p.m.
25. 3.30 p.m.	**26.** 5.45 p.m.	**27.** 7.20 p.m.	**28.** 9.40 p.m.
29. 11.35 p.m.	**30.** 12.10 a.m.		

P. 15   EX. 12

1.	12.00	**2.**	24.00	**3.**	01.00	**4.**	03.00
5.	06.00	**6.**	08.00	**7.**	10.00	**8.**	13.00
9.	15.00	**10.**	18.00	**11.**	20.00	**12.**	22.00
13.	04.15	**14.**	07.30	**15.**	09.45	**16.**	11.20
17.	14.10	**18.**	16.25	**19.**	17.35	**20.**	19.40
21.	21.20	**22.**	22.50	**23.**	23.12	**24.**	23.55

P. 16   EX. 13

1.	30	**2.**	31	**3.**	28	**4.**	30	**5.**	31
6.	31	**7.**	31	**8.**	31	**9.**	30	**10.**	31
11.	30	**12.**	31						

P. 17   EX. 14

1.	51	**2.**	120	**3.**	69	**4.**	92	**5.**	76	**6.**	92
7.	81	**8.**	92	**9.**	86	**10.**	116	**11.**	32	**12.**	60
13.	62	**14.**	62	**15.**	62	**16.**	62	**17.**	63	**18.**	62
19.	62	**20.**	62	**21.**	37	**22.**	52	**23.**	50	**24.**	68
25.	66	**26.**	60	**27.**	86	**28.**	100	**29.**	70	**30.**	113

P. 19   EX. 15

1.	6	**2.**	8	**3.**	10	**4.**	12	**5.**	16	**6.**	20
7.	25	**8.**	32	**9.**	40	**10.**	50	**11.**	100	**12.**	1000
13.	4	**14.**	6	**15.**	8	**16.**	10	**17.**	14	**18.**	18
19.	30	**20.**	60								

P. 20   EX. 16

1.	3	**2.**	6	**3.**	10	**4.**	12	**5.**	5	**6.**	8
7.	7	**8.**	9	**9.**	4	**10.**	11	**11.**	8	**12.**	12
13.	16	**14.**	20	**15.**	24	**16.**	6	**17.**	10	**18.**	14
19.	18	**20.**	22	**21.**	5	**22.**	7	**23.**	9	**24.**	11
25.	13	**26.**	15	**27.**	17	**28.**	19	**29.**	21	**30.**	23

P. 20   EX. 17

1.	16	**2.**	32	**3.**	48	**4.**	12	**5.**	20	**6.**	28
7.	34	**8.**	42	**9.**	14	**10.**	22	**11.**	9	**12.**	19
13.	29	**14.**	39	**15.**	46	**16.**	36	**17.**	26	**18.**	38
19.	11	**20.**	37	**21.**	16	**22.**	48	**23.**	80	**24.**	8

25.	4	26.	12	27.	2	28.	6	29.	10	30.	14
31.	40	32.	36	33.	42	34.	54	35.	62	36.	76
37.	69	38.	87	39.	91	40.	95				

P. 21 EX. 18

1.	6	2.	12	3.	18	4.	4	5.	11	6.	16
7.	7	8.	14	9.	10	10.	17	11.	12	12.	24
13.	36	14.	48	15.	60	16.	4	17.	8	18.	16
19.	32	20.	40	21.	56	22.	64	23.	20	24.	28
25.	68	26.	13	27.	29	28.	41	29.	55	30.	67
31.	23	32.	35	33.	71	34.	47	35.	59		

P. 22 EX. 19

1.	15	2.	20	3.	25	4.	6	5.	12	6.	21
7.	17	8.	13	9.	9	10.	28	11.	8	12.	12
13.	19	14.	23	15.	26	16.	10	17.	20	18.	30
19.	40	20.	50	21.	60	22.	12	23.	24	24.	36
25.	48	26.	52	27.	44	28.	38	29.	26	30.	18
31.	11	32.	13	33.	17	34.	19	35.	23	36.	33
37.	43	38.	47	39.	57	40.	59				

P. 22 EX. 20

1. $\frac{2}{4}$ 2. $\frac{4}{6}$ 3. $\frac{4}{5}$ 4. $\frac{4}{6}$ 5. $\frac{2}{4}$ 6. $\frac{4}{10}$

7. $\frac{1}{2}$ 8. $\frac{1}{4}$ 9. $\frac{3}{4}$ 10. $\frac{1}{2}$ 11. $\frac{1}{3}$ 12. $\frac{1}{4}$

13. $\frac{2}{10}$ 14. $\frac{6}{15}$ 15. $\frac{12}{20}$ 16. $\frac{5}{20}$ 17. $\frac{12}{24}$ 18. $\frac{21}{28}$

19. $\frac{1}{7}$ 20. $\frac{3}{8}$

P. 23 EX. 21

1. $\frac{1}{2}$ 2. $\frac{2}{3}$ 3. $\frac{1}{3}$ 4. $\frac{2}{5}$ 5. $\frac{1}{4}$ 6. $\frac{1}{2}$

7. $\frac{1}{3}$ 8. $\frac{1}{4}$ 9. $\frac{1}{5}$ 10. $\frac{1}{6}$ 11. $\frac{1}{2}$ 12. $\frac{1}{3}$

13. $\frac{1}{4}$ 14. $\frac{1}{5}$ 15. $\frac{1}{6}$ 16. $\frac{1}{2}$ 17. $\frac{2}{3}$ 18. $\frac{1}{3}$

19. $\frac{2}{5}$ 20. $\frac{1}{4}$ 21. $\frac{1}{2}$ 22. $\frac{1}{3}$ 23. $\frac{1}{4}$ 24. $\frac{1}{5}$

25. $\frac{1}{6}$ 26. $\frac{1}{2}$ 27. $\frac{1}{3}$ 28. $\frac{1}{4}$ 29. $\frac{1}{5}$ 30. $\frac{1}{6}$

31. $\frac{1}{2}$ 32. $\frac{2}{3}$ 33. $\frac{1}{3}$ 34. $\frac{2}{5}$ 35. $\frac{1}{4}$ 36. $\frac{1}{2}$

37. $\frac{1}{3}$ 38. $\frac{1}{4}$ 39. $\frac{1}{5}$ 40. $\frac{1}{6}$ 41. $\frac{1}{2}$ 42. $\frac{1}{3}$

43. $\frac{1}{4}$ 44. $\frac{1}{5}$ 45. $\frac{1}{6}$ 46. $\frac{1}{2}$ 47. $\frac{2}{3}$ 48. $\frac{1}{3}$

49. $\frac{2}{5}$ 50. $\frac{1}{4}$ 51. $\frac{1}{2}$ 52. $\frac{1}{3}$ 53. $\frac{1}{4}$ 54. $\frac{1}{5}$

55. $\frac{1}{6}$ 56. $\frac{3}{4}$ 57. $\frac{3}{8}$ 58. $\frac{4}{5}$ 59. $\frac{5}{8}$ 60. $\frac{3}{5}$

61. $\frac{7}{10}$ 62. $\frac{7}{8}$ 63. $\frac{5}{12}$ 64. $\frac{5}{9}$ 65. $\frac{6}{7}$ 66. $\frac{7}{9}$

67. $\frac{4}{5}$ 68. $\frac{7}{9}$ 69. $\frac{1}{3}$ 70. $\frac{8}{15}$ 71. $\frac{7}{12}$ 72. $\frac{1}{3}$

73. $\frac{3}{4}$ 74. $\frac{9}{16}$ 75. $\frac{4}{5}$ 76. $\frac{3}{5}$ 77. $\frac{3}{4}$ 78. $\frac{3}{4}$

79. $\frac{4}{5}$ 80. $\frac{1}{4}$

P. 25 EX. 22

1. $\frac{1}{2}, \frac{3}{5}, \frac{7}{10}$ 2. $\frac{1}{4}, \frac{3}{10}, \frac{2}{5}$ 3. $\frac{1}{2}, \frac{5}{8}, \frac{3}{4}$ 4. $\frac{23}{32}, \frac{3}{4}, \frac{13}{16}$

5. $\frac{1}{4}, \frac{3}{8}, \frac{7}{16}$ 6. $\frac{4}{9}, \frac{1}{2}, \frac{2}{3}$ 7. $\frac{8}{15}, \frac{3}{5}, \frac{2}{3}$ 8. $\frac{2}{9}, \frac{5}{18}, \frac{1}{3}$

9. $\frac{7}{9}, \frac{29}{36}, \frac{5}{6}$ 10. $\frac{19}{30}, \frac{13}{20}, \frac{7}{10}$ 11. $\frac{2}{3}, \frac{7}{9}, \frac{5}{6}$ 12. $\frac{5}{6}, \frac{8}{9}, \frac{11}{12}$

13. $\frac{3}{5}, \frac{2}{3}, \frac{5}{6}$ 14. $\frac{5}{21}, \frac{2}{7}, \frac{1}{3}$ 15. $\frac{11}{15}, \frac{23}{30}, \frac{4}{5}$ 16. $\frac{3}{4}, \frac{5}{6}, \frac{7}{8}$

17. $\frac{3}{4}, \frac{7}{9}, \frac{5}{6}$ 18. $\frac{5}{6}, \frac{7}{8}, \frac{11}{12}$ 19. $\frac{4}{5}, \frac{17}{20}, \frac{7}{8}$ 20. $\frac{7}{9}, \frac{22}{27}, \frac{5}{6}$

21. $\frac{13}{24}, \frac{7}{12}, \frac{5}{8}$ 22. $\frac{7}{22}, \frac{4}{11}, \frac{13}{33}$ 23. $\frac{17}{20}, \frac{13}{15}, \frac{9}{10}$ 24. $\frac{8}{15}, \frac{5}{9}, \frac{3}{5}$

1. $1\frac{1}{2}$ 2. $1\frac{2}{3}$ 3. $1\frac{3}{5}$ 4. $3\frac{1}{2}$ 5. $2\frac{1}{3}$ 6. $1\frac{3}{4}$

7. $1\frac{2}{5}$ 8. $1\frac{1}{4}$ 9. $2\frac{2}{3}$ 10. $1\frac{1}{3}$ 11. $4\frac{1}{2}$ 12. 3

13. $2\frac{1}{4}$ 14. $1\frac{4}{5}$ 15. $1\frac{1}{2}$ 16. 5 17. $3\frac{1}{3}$ 18. $2\frac{1}{2}$

19. 2 20. $1\frac{2}{3}$ 21. $3\frac{2}{3}$ 22. $2\frac{3}{4}$ 23. $2\frac{1}{5}$ 24. $2\frac{2}{5}$

25. $1\frac{5}{7}$ 26. $1\frac{5}{8}$ 27. $1\frac{2}{3}$ 28. $1\frac{7}{10}$ 29. $1\frac{8}{11}$ 30. $1\frac{3}{4}$

31. $2\frac{1}{3}$ 32. $3\frac{1}{5}$ 33. $3\frac{1}{7}$ 34. $2\frac{3}{5}$ 35. 2 36. $3\frac{5}{6}$

37. $5\frac{3}{4}$ 38. $9\frac{1}{2}$ 39. 5 40. 5 41. $3\frac{5}{9}$ 42. $3\frac{2}{9}$

43. 2 44. $2\frac{2}{9}$ 45. $4\frac{2}{7}$

1. $\frac{3}{2}$ 2. $\frac{5}{4}$ 3. $\frac{7}{4}$ 4. $\frac{4}{3}$ 5. $\frac{5}{3}$ 6. $\frac{9}{8}$

7. $\frac{11}{8}$ 8. $\frac{13}{8}$ 9. $\frac{15}{8}$ 10. $\frac{9}{4}$ 11. $\frac{19}{8}$ 12. $\frac{8}{3}$

13. $\frac{15}{4}$ 14. $\frac{29}{8}$ 15. $\frac{19}{6}$ 16. $\frac{17}{4}$ 17. $\frac{22}{5}$ 18. $\frac{29}{6}$

19. $\frac{31}{7}$ 20. $\frac{47}{10}$ 21. $\frac{11}{2}$ 22. $\frac{16}{3}$ 23. $\frac{21}{4}$ 24. $\frac{28}{5}$

25. $\frac{45}{8}$ 26. $\frac{20}{3}$ 27. $\frac{27}{4}$ 28. $\frac{32}{5}$ 29. $\frac{51}{8}$ 30. $\frac{41}{6}$

31. $\frac{23}{3}$ 32. $\frac{35}{4}$ 33. $\frac{38}{5}$ 34. $\frac{67}{8}$ 35. $\frac{61}{8}$ 36. $\frac{33}{4}$

37. $\frac{15}{2}$ 38. $\frac{42}{5}$ 39. $\frac{31}{4}$ 40. $\frac{26}{3}$ 41. $\frac{48}{5}$ 42. $\frac{165}{16}$

43. $\frac{25}{2}$ 44. $\frac{47}{4}$ 45. $\frac{44}{3}$

P. 30   Bisecting angle PQR

Angles PQS and RQS should be equal. QS bisects angle PQR.

P. 31   Bisecting straight line PQ

PT and QT should be equal. RS bisects PQ. Each of the angles at T is 90°.

P. 32   Drawing TX perpendicular to PQ

Angles RXT and SXT are each 90°.

P. 32   Drawing AB perpendicular to CD

Angles PBA and QBA are each 90°.

P. 35   EX. 27

1   XY and PQ are parallel to each other.
2   KL and PQ are parallel to each other.
3   The new figure is a square. You cannot construct a figure inside the original square, 2 cm from the sides.

P. 39   Prime numbers

1   Every 4th number after 4 will have been crossed out with the 2nd numbers after 2. (2 is a factor of 4.) This is also true for every 6th number after 6 and every 8th number after 8.
2   The next set of numbers after 13 will be every 17th number after 17.

Prime numbers 1, 2, 3, 5, 7, 11, 13, 17, 19, 23, 29, 31, 37, 41, 43, 47, 49, 53, 59, 61, 67, 71, 73, 79, 83, 89, 91, 97.

P. 39   Questions

(1) 3^4 (2) 4^3 (3) 5^2 (4) 6^5 (5) 2^4 (6) 3^1

P. 40   EX. 29

1. $2^2 \times 3^2$	2. $3^3 \times 5^2$	3. $2^2 \times 3^2 \times 5^2$	4. $2^3 \times 3^2$
5. $2^2 \times 3^3$	6. $3^2 \times 5^3$	7. $3^3 \times 5^2$	8. $2^2 \times 3^3 \times 5^3$
9. $2^3 \times 3 \times 5$	10. $2 \times 3^3 \times 5^2$	11. $2^2 \times 3^2$	12. $2^2 \times 5^2$
13. $2^2 \times 3^2 \times 5^2$	14. $2^3 \times 3^2 \times 5^2$	15. $2^6 \times 3^2$	16. $2^7 \times 3^3$
17. $2^3 \times 3^6$	18. $2^4 \times 3^4 \times 5^2$	19. $2^3 \times 3^3 \times 5^3$	20. $2^3 \times 3^3 \times 5^2$

P. 41   EX. 30

1. 2^3	2. 3^2	3. 5^2	4. $2^2 \times 3$
5. 2^4	6. $2^2 \times 5$	7. $2 \times 3 \times 5$	8. $2^3 \times 5$
9. 2×5^2	10. 3×5^2	11. $2^2 \times 3^2$	12. $2^4 \times 3$
13. $2^2 \times 3 \times 5$	14. $2^3 \times 3^2$	15. $2^3 \times 7$	16. 2^6
17. $2 \times 5 \times 7$	18. $2^4 \times 5$	19. $2^2 \times 3 \times 7$	20. $2^5 \times 3$
21. $2^2 \times 5^2$	22. $2^3 \times 3 \times 5$	23. $2^4 \times 3^2$	24. $2^3 \times 3^3$

25.	$3^2 \times 5^2$	26.	5^3		27.	$2^3 \times 17$	28.	2^8
29.	$2^2 \times 3 \times 5^2$	30.	$2^3 \times 3^2 \times 5$	31.	$2 \times 3^2 \times 5^2$	32.	$2^2 \times 3^3 \times 5$
33.	$2^4 \times 3^3$	34.	$2^5 \times 3^3$		35.	$2^4 \times 3 \times 5^2$	36.	$2^6 \times 5^2$
37.	$2^5 \times 5 \times 11$	38.	$2^4 \times 3 \times 5^3$	39.	$2^4 \times 3^3 \times 5^2$	40.	$3^2 \times 5^2 \times 7^2$

P. 41   EX. 31

1.	2	2.	3	3.	2	4.	4	5.	3
6.	4	7.	4	8.	3	9.	3	10.	5
11.	5	12.	12	13.	3	14.	6	15.	10
16.	5	17.	8	18.	16	19.	7	20.	9

P. 42   EX. 32

1.	12	2.	12	3.	12	4.	8	5.	12
6.	16	7.	16	8.	18	9.	10	10.	15
11.	24	12.	30	13.	36	14.	24	15.	36
16.	30	17.	36	18.	42	19.	72	20.	36
21.	180	22.	60	23.	48	24.	48	25.	60
26.	42	27.	120	28.	210				

P. 43   EX. 33

1.	50p	2.	10p	3.	30 min	4.	12 h	5.	20p
6.	25p	7.	12 min	8.	6 h	9.	£3	10.	£2
11.	75p	12.	60p	13.	45°	14.	60°	15.	4 days
16.	8 mths	17.	39 wks	18.	50 min	19.	£1.50	20.	74p

P. 43   EX. 34

1.	$\frac{1}{4}$	2.	$\frac{1}{2}$	3.	$\frac{3}{4}$	4.	$\frac{1}{10}$	5.	$\frac{1}{5}$
6.	$\frac{2}{5}$	7.	$\frac{1}{10}$	8.	$\frac{1}{4}$	9.	$\frac{1}{2}$	10.	$\frac{3}{5}$
11.	$\frac{1}{3}$	12.	$\frac{1}{2}$	13.	$\frac{1}{3}$	14.	$\frac{1}{4}$	15.	$\frac{2}{7}$
16.	$\frac{2}{3}$	17.	$\frac{3}{4}$	18.	$\frac{5}{6}$	19.	$\frac{3}{5}$	20.	$\frac{2}{5}$

1. $\dfrac{3}{4}$　2. $\dfrac{5}{6}$　3. $\dfrac{7}{10}$　4. $\dfrac{7}{12}$　5. $\dfrac{8}{15}$　6. $\dfrac{9}{20}$

7. $1\dfrac{1}{6}$　8. $\dfrac{11}{12}$　9. $1\dfrac{1}{15}$　10. $1\dfrac{1}{12}$　11. $1\dfrac{5}{12}$　12. $\dfrac{11}{15}$

13. $\dfrac{14}{15}$　14. $1\dfrac{7}{15}$　15. $1\dfrac{1}{4}$　16. $\dfrac{9}{10}$　17. $\dfrac{5}{8}$　18. $\dfrac{3}{8}$

19. $\dfrac{11}{24}$　20. $\dfrac{13}{40}$　21. $1\dfrac{1}{24}$　22. $1\dfrac{9}{40}$　23. $1\dfrac{7}{20}$　24. $1\dfrac{11}{40}$

25. $1\dfrac{1}{12}$　26. $1\dfrac{5}{12}$　27. $1\dfrac{7}{12}$　28. $1\dfrac{11}{12}$　29. $\dfrac{3}{4}$　30. $1\dfrac{1}{12}$

31. $1\dfrac{1}{4}$　32. $1\dfrac{5}{12}$　33. $1\dfrac{3}{4}$　34. $2\dfrac{1}{4}$　35. $\dfrac{7}{16}$　36. $\dfrac{13}{16}$

37. $1\dfrac{3}{16}$　38. $1\dfrac{5}{16}$　39. $1\dfrac{13}{16}$　40. $1\dfrac{35}{36}$　41. $1\dfrac{17}{36}$　42. $\dfrac{23}{24}$

43. $2\dfrac{1}{24}$　44. $1\dfrac{29}{36}$　45. $1\dfrac{13}{24}$　46. $1\dfrac{29}{66}$　47. $1\dfrac{37}{60}$　48. $1\dfrac{11}{12}$

1. $\dfrac{1}{4}$　2. $\dfrac{1}{6}$　3. $\dfrac{3}{10}$　4. $\dfrac{1}{12}$　5. $\dfrac{2}{15}$　6. $\dfrac{1}{20}$

7. $\dfrac{1}{6}$　8. $\dfrac{5}{12}$　9. $\dfrac{4}{15}$　10. $\dfrac{5}{12}$　11. $\dfrac{1}{12}$　12. $\dfrac{1}{15}$

13. $\dfrac{4}{15}$　14. $\dfrac{2}{15}$　15. $\dfrac{1}{4}$　16. $\dfrac{1}{10}$　17. $\dfrac{3}{8}$　18. $\dfrac{1}{8}$

19. $\dfrac{5}{24}$　20. $\dfrac{3}{40}$　21. $\dfrac{7}{24}$　22. $\dfrac{1}{40}$　23. $\dfrac{3}{20}$　24. $\dfrac{19}{40}$

25. $\dfrac{1}{6}$　26. $\dfrac{1}{12}$　27. $\dfrac{1}{30}$　28. $\dfrac{1}{20}$　29. $\dfrac{1}{30}$　30. $\dfrac{2}{15}$

31. $\dfrac{1}{12}$　32. $\dfrac{5}{24}$　33. $\dfrac{7}{36}$　34. $\dfrac{1}{36}$　35. $\dfrac{1}{24}$　36. $\dfrac{1}{24}$

37. $\dfrac{1}{36}$　38. $\dfrac{5}{28}$　39. $\dfrac{1}{35}$　40. $\dfrac{1}{24}$　41. $\dfrac{13}{60}$　42. $\dfrac{1}{60}$

43. $\dfrac{7}{60}$　44. $\dfrac{1}{66}$　45. $\dfrac{3}{56}$　46. $\dfrac{1}{42}$　47. $\dfrac{1}{30}$　48. $\dfrac{1}{72}$

1. $5\dfrac{11}{12}$　2. $5\dfrac{5}{12}$　3. $6\dfrac{7}{24}$　4. $5\dfrac{13}{16}$　5. $3\dfrac{5}{6}$　6. $6\dfrac{11}{36}$

7. $4\frac{7}{8}$ 8. $4\frac{23}{66}$ 9. $5\frac{29}{36}$ 10. $6\frac{7}{8}$ 11. $6\frac{9}{20}$ 12. $5\frac{11}{12}$

13. $5\frac{35}{36}$ 14. $5\frac{5}{36}$ 15. $5\frac{31}{36}$ 16. $4\frac{3}{4}$ 17. $5\frac{11}{12}$ 18. $5\frac{3}{4}$

19. $10\frac{3}{8}$ 20. $8\frac{7}{60}$ 21. $9\frac{31}{66}$ 22. $7\frac{5}{6}$ 23. $7\frac{47}{54}$ 24. $4\frac{5}{48}$

P. 50   EX. 38

1. $1\frac{1}{6}$ 2. $2\frac{1}{8}$ 3. $\frac{11}{15}$ 4. $2\frac{1}{12}$ 5. $2\frac{1}{10}$ 6. $2\frac{1}{30}$

7. $\frac{5}{12}$ 8. $1\frac{19}{36}$ 9. $\frac{17}{24}$ 10. $1\frac{1}{24}$ 11. $\frac{35}{36}$ 12. $2\frac{7}{30}$

13. $\frac{23}{60}$ 14. $\frac{29}{36}$ 15. $1\frac{25}{28}$ 16. $\frac{35}{36}$ 17. $\frac{13}{18}$ 18. $1\frac{31}{42}$

19. $1\frac{5}{24}$ 20. $\frac{44}{45}$ 21. $2\frac{47}{48}$ 22. $1\frac{25}{33}$ 23. $\frac{5}{12}$ 24. $2\frac{1}{3}$

P. 51   EX. 39

1. $\frac{7}{12}$ 2. $2\frac{1}{3}$ 3. $\frac{3}{5}$ 4. $\frac{7}{8}$ 5. $5\frac{1}{12}$ 6. $1\frac{1}{18}$

7. $\frac{2}{3}$ 8. $1\frac{3}{8}$ 9. $\frac{1}{6}$ 10. $3\frac{3}{4}$ 11. $\frac{5}{12}$ 12. $1\frac{8}{15}$

13. $1\frac{29}{30}$ 14. 1 15. $1\frac{7}{36}$

P. 54   Equilateral triangle **Q.5** Each angle equals $60°$
P. 55   Second method **Q.4** AC = BC = 6 cm. Angle ACB = $60°$

P. 56   Square

(A)   CD = 6 cm   **(B)** Angles ADC and BCD each equal $90°$

P. 57   Pentagon

(A)   ED = CD = 6 cm   (B) Angle EDC = $108°$

P. 58   Hexagon

(1)   DE = 6 cm   **(2)** Angles CDE and FED each equal $120°$

(1) FE = 5 cm   **(2)** Angles GFE and DEF each equal 135°
(3) 6 cm sides would make the diagram too large for the exercise book.

P. 61

Name of diagram	Number of sides	Number of interior angles	Size of each interior angle	Sum of all the interior angles
Equilateral triangle	3	3	60°	180°
Square	4	4	90°	360°
Regular pentagon	5	5	108°	540°
Regular hexagon	6	6	120°	720°
Regular octagon	8	8	135°	1080°

The sum of all the interior angles is not the same answer each time.

P. 61

| Name of diagram | Number of sides | Multiply number of sides by 2 | Multiply 90° by answer **A** | Subtract 360° from answer **B** |
		A	**B**	**C**
Equilateral triangle	3	6	540°	180°
Square	4	8	720°	360°
Regular pentagon	5	10	900°	540°
Regular hexagon	6	12	1080°	720°
Regular octagon	8	16	1440°	1080°

The answers in column C agree with the answers for 'the sum of the interior angles' in the previous table.

P. 62

Name of diagram	Number of sides N	2 N	2 N − 4	(2 N − 4) X 90°
Equilateral triangle	3	6	2	180°
Square	4	8	4	360°
Regular pentagon	5	10	6	540°
Regular hexagon	6	12	8	720°
Regular octagon	8	16	12	1080°

P. 64

Name of diagram	Number of sides	Number of exterior angles	Size of each exterior angle	Sum of all the exterior angles
Equilateral triangle	3	3	120°	360°
Square	4	4	90°	360°
Regular pentagon	5	5	72°	360°
Regular hexagon	6	6	60°	360°
Regular octagon	8	8	45°	360°

P. 66 Inscribed equilateral triangle

1 AB = BC = CA = approx 5·2 cm
2 Angles ABC, BCA and CAB each equal 60°

P. 66 Inscribed square

1 AB = BC = CD = DA = approx 4·2 cm
2 Angles ABC, BCD, CDA and DAB each equal 90°

Name of regular polygon	Number of equal sides	Number of equal angles at centre of circle	Size of each angle at centre of circle
Equilateral triangle	3	3	$360° \div 3 = 120°$
Square	4	4	$360° \div 4 = 90°$
Pentagon	5	5	$360° \div 5 = 72°$
Hexagon	6	6	$360° \div 6 = 60°$
Octagon	8	8	$360° \div 8 = 45°$

P. 68   EX. 40

1. $\frac{2}{3}$ 2. $2\frac{5}{8}$ 3. $3\frac{2}{5}$ 4. $2\frac{5}{12}$ 5. $4\frac{7}{12}$ 6. $2\frac{11}{12}$

7. $2\frac{13}{14}$ 8. $2\frac{7}{24}$ 9. $1\frac{7}{8}$ 10. $1\frac{9}{16}$ 11. $3\frac{5}{18}$ 12. $2\frac{9}{20}$

13. $3\frac{1}{5}$ 14. $3\frac{1}{12}$ 15. $2\frac{1}{9}$ 16. $2\frac{14}{15}$ 17. $2\frac{5}{24}$ 18. $2\frac{31}{36}$

19. $1\frac{3}{4}$ 20. $\frac{7}{8}$ 21. $1\frac{9}{28}$ 22. $1\frac{11}{60}$ 23. $1\frac{7}{9}$ 24. $\frac{29}{56}$

P. 71   EX. 41

1. 4   2. 3   3. 2   4. 2   5. $\frac{1}{6}$ 6. $\frac{1}{8}$

7. $\frac{1}{10}$ 8. $\frac{1}{12}$ 9. $\frac{1}{12}$ 10. $\frac{1}{15}$ 11. $\frac{1}{18}$ 12. $\frac{1}{24}$

13. $\frac{1}{20}$ 14. $\frac{1}{24}$ 15. $\frac{1}{32}$ 16. $\frac{1}{40}$ 17. 2   18. 3

19. 4   20. 5   21. 4   22. 9   23. 12   24. 10

25. $1\frac{1}{3}$ 26. $1\frac{1}{4}$ 27. $1\frac{1}{5}$ 28. $1\frac{1}{2}$ 29. $3\frac{1}{3}$ 30. $5\frac{1}{4}$

31. $7\frac{1}{5}$ 32. 10   33. $\frac{3}{8}$ 34. $\frac{3}{10}$ 35. $\frac{2}{5}$ 36. $\frac{3}{7}$

37. $\frac{1}{4}$ 38. $\frac{1}{5}$ 39. $\frac{2}{9}$ 40. $\frac{4}{15}$ 41. $\frac{2}{5}$ 42. $\frac{5}{12}$

43. $\frac{5}{8}$ 44. $\frac{3}{5}$ 45. $\frac{5}{9}$ 46. $\frac{3}{10}$ 47. $\frac{3}{10}$ 48. $\frac{5}{16}$

49. $\frac{1}{2}$ 50. $\frac{2}{3}$ 51. $\frac{1}{4}$ 52. $\frac{3}{4}$ 53. $\frac{1}{8}$ 54. $\frac{1}{2}$

55. $\frac{1}{6}$ **56.** $\frac{1}{6}$ **57.** $\frac{1}{2}$ **58.** $\frac{1}{8}$ **59.** $\frac{1}{3}$ **60.** $\frac{2}{7}$

61. $\frac{1}{4}$ **62.** $\frac{1}{3}$ **63.** $\frac{1}{4}$ **64.** $\frac{2}{3}$ **65.** $\frac{1}{8}$ **66.** $\frac{1}{3}$

67. $\frac{3}{5}$ **68.** $\frac{1}{5}$

P. 72 EX. 42

1. 1 **2.** 1 **3.** 1 **4.** $3\frac{1}{2}$ **5.** 8 **6.** $5\frac{1}{2}$

7. 2 **8.** $4\frac{1}{2}$ **9.** 1 **10.** 2 **11.** 1 **12.** 1

13. 2 **14.** 2 **15.** 2 **16.** 2 **17.** 4 **18.** 6

19. $4\frac{1}{2}$ **20.** 6 **21.** 6 **22.** 3 **23.** 4 **24.** 4

25. $1\frac{1}{2}$ **26.** $2\frac{1}{4}$ **27.** $4\frac{1}{2}$ **28.** $3\frac{3}{5}$ **29.** 2 **30.** 18

31. 4 **32.** 4 **33.** $3\frac{1}{2}$ **34.** $1\frac{1}{3}$ **35.** 18 **36.** $1\frac{1}{3}$

P. 74 EX. 43

1. $\frac{2}{5}$ **2.** $\frac{3}{4}$ **3.** $\frac{4}{5}$ **4.** $\frac{5}{7}$ **5.** $\frac{1}{3}$ **6.** $\frac{1}{2}$

7. $\frac{1}{3}$ **8.** $\frac{3}{4}$ **9.** $\frac{2}{3}$ **10.** $\frac{2}{3}$ **11.** $\frac{4}{5}$ **12.** $\frac{3}{4}$

13. $\frac{8}{9}$ **14.** $\frac{2}{3}$ **15.** $\frac{1}{3}$ **16.** $\frac{3}{4}$ **17.** $\frac{2}{3}$ **18.** $\frac{3}{4}$

19. $\frac{3}{5}$ **20.** $\frac{3}{5}$ **21.** 8 **22.** 6 **23.** 12 **24.** 20

25. 9 **26.** 10 **27.** 16 **28.** 10 **29.** $7\frac{1}{2}$ **30.** $2\frac{2}{3}$

31. 5 **32.** $4\frac{4}{5}$ **33.** 9 **34.** $7\frac{1}{2}$ **35.** $10\frac{2}{3}$ **36.** 12

37. $\frac{1}{4}$ **38.** $\frac{1}{9}$ **39.** $\frac{1}{16}$ **40.** $\frac{1}{25}$ **41.** $\frac{1}{6}$ **42.** $\frac{1}{12}$

43. $\frac{1}{20}$ **44.** $\frac{1}{30}$ **45.** $\frac{1}{3}$ **46.** $\frac{1}{4}$ **47.** $\frac{1}{5}$ **48.** $\frac{1}{6}$

49. $\frac{1}{8}$ **50.** $\frac{4}{21}$ **51.** $\frac{1}{9}$ **52.** $\frac{6}{55}$ **53.** 1 **54.** 1

55. 1 **56.** 1 **57.** $1\frac{1}{2}$ **58.** $1\frac{1}{3}$ **59.** $1\frac{1}{4}$ **60.** $1\frac{1}{5}$

61. $\frac{3}{4}$	62. $\frac{8}{9}$	63. $\frac{15}{16}$	64. $\frac{24}{25}$	65. $\frac{4}{5}$	66. $1\frac{1}{5}$
67. $1\frac{1}{15}$	68. $1\frac{1}{24}$	69. $\frac{1}{2}$	70. 2	71. $\frac{3}{4}$	72. $3\frac{1}{4}$
73. $\frac{2}{3}$	74. $1\frac{1}{2}$	75. $\frac{3}{4}$	76. $1\frac{1}{3}$	77. 2	78. 4
79. $1\frac{1}{9}$	80. $1\frac{1}{3}$	81. $2\frac{1}{2}$	82. $1\frac{1}{2}$	83. $1\frac{1}{2}$	84. $\frac{3}{4}$

P. 78 EX. 44

1. 432 mm	2. 345 mm	3. 683 mm	4. 3720 mm
5. 2704 mm	6. 5068 mm	7. 40302 mm	8. 2050700 mm
9. 5002006 mm	10. 235 m	11. 486 m	12. 793 m
13. 3207 m	14. 4032 m	15. 5620 m	16. 5·63 m
17. 7·408 m	18. 9·065 m	19. 30·709 m	20. 506·07 m
21. 8004·003 m	22. 5·34 km	23. 4·062 km	
24. 8·506 km	25. 3·0402 km	26. 2·0046 km	
27. 6·008005 km	28. 0·675 km	29. 0·030507 km	
30. 0·009064 km			

P. 79

Question 1   decametres.   **Question 2**   667 dam.

P. 80   EX. 45

1. 10·62 km	2. 11·9 km	3. 10·22 km	4. 8·34 m
5. 13·14 m	6. 9·77 m	7. 8·54 cm	8. 15·62 cm
9. 8·2 cm	10. 10·916 m	11. 13·294 m	12. 8·912 km
13. 14·888 m	14. 20·813 hm	15. 14·885244 km	

P. 80   EX. 46

1. 323 dam	2. 199 dam	3. 279 dam	4. 328 cm
5. 193 cm	6. 267 cm	7. 79 mm	8. 347 mm
9. 5 mm	10. 67 cm	11. 6602 mm	12. 465371 cm
13. 4413 cm	14. 1683315 cm	15. 26336 mm	

P. 82   EX. 47

1. 22·4 m	2. 28·2 km	3. 41·3 mm	4. 29·6 cm
5. 28·5 m	6. 42 km	7. 28·08 mm	8. 18·72 cm
9. 31·92 m	10. 2·3 cm	11. 2·1 mm	12. 1·8 km

13.	1·4 km	14.	1·9 m	15.	2·6 cm	16.	2·1 mm
17.	1·2 km	18.	2·2 m	19.	38·16 km	20.	68·16 m
21.	99·66 cm	22.	89·76 mm	23.	58·86 km	24.	68·75 m
25.	18·368 cm	26.	29·952 cm	27.	3·2 km	28.	1·8 cm
29.	3·6 m	30.	4·8 mm	31.	3·2 cm	32.	2·3 km
33.	263·68 mm	34.	250 cm	35.	0·27343 km	36.	0·64 cm

P. 84   EX. 48

1.	5 cm^2	2.	7 cm^2	3.	5 cm^2	4.	8 cm^2	5.	7 cm^2
6.	5 cm^2	7.	7 cm^2	8.	6 cm^2	9.	6 cm^2	10.	7 cm^2
11.	6 cm^2	12.	8 cm^2	13.	6 cm^2	14.	12 cm^2	15.	20 cm^2

P. 89   EX. 49

1.	16 cm^2	2.	200 cm^2	3.	128 cm^2	4.	63 m^2
5.	84 cm^2	6.	128 m^2	7.	198 km^2	8.	450 km^2
9.	432 km^2	10.	300 mm^2	11.	720 mm^2	12.	1536 mm^2
13.	15 cm^2	14.	26 mm^2	15.	39 m^2	16.	11 km^2
17.	27 m^2	18.	27 cm^2	19.	22 mm^2	20.	17 cm^2
21.	27 km^2	22.	5 m^2	23.	22 mm^2	24.	$17·6 \text{ cm}^2$
25.	17 km^2	26.	28 m^2	27.	10 cm^2	28.	21 km^2
29.	42 m^2	30.	36 cm^2	31.	$14·28 \text{ km}^2$	32.	$20·72 \text{ mm}^2$
33.	$19·2 \text{ cm}^2$	34.	$53·256 \text{ m}^2$	35.	$59·04 \text{ mm}^2$	36.	$188·6 \text{ km}^2$

P. 91   EX. 50

1.	12 cm	2.	12 cm	3.	12 cm	4.	14 cm	5.	16 cm
6.	12 cm	7.	16 cm	8.	14 cm	9.	14 cm	10.	16 cm
11.	14 cm	12.	16 cm	13.	10 cm	14.	14 cm	15.	18 cm

P. 91   EX. 51

1.	18 cm ; 11 cm^2	2.	18 cm ; 12 cm^2	3.	16 cm ; 13 cm^2
4.	24 cm ; 20 cm^2	5.	24 cm ; 26 cm^2	6.	34 cm ; 36 cm^2
7.	40 cm ; 48 cm^2	8.	30 cm ; 29 cm^2	9.	44 cm ; 40 cm^2
10.	44 cm ; 52 cm^2	11.	36 cm ; 44 cm^2	12.	26 cm ; $17½ \text{ cm}^2$

P. 94   EX. 52

1.	8 cm^3	2.	24 cm^3	3.	12 cm^3	4.	18 cm^3
5.	27 m^3	6.	36 mm^3	7.	30 cm^3	8.	50 cm^3
9.	80 mm^3	10.	64 cm^3	11.	150 cm^3	12.	200 cm^3
13.	50 m^3	14.	54 m^3	15.	125 mm^3	16.	1000 cm^3

17.	180 m³	18.	210 m³	19.	100 cm³	20.	210 cm³
21.	32·4 m³	22.	54 cm³	23.	108 cm³	24.	201·6 cm³
25.	133 m³	26.	223·2 m³	27.	77·91 cm³	28.	33·6 cm³
29.	225·99 cm³	30.	2430 cm³				

P. 95   EX. 53

No.	L (cm)	B (cm)	A (cm²)	No.	L (cm)	B (cm)	A (cm²)	No.	L (cm)	B (cm)	A (cm²)
1			12	11			100	21			90
2		3		12	24			22		6½	
3			20	13		16		23		4	
4		6		14			360	24	9⅓		
5			·30	15		4		25			42
6	6			16			396	26		8⅔	
7		5		17	20			27	5		
8			63	18		20		28		8	
9	7			19			720	29			86
10		12		20	32			30	7½		

P. 95   EX. 54

No.	L (cm)	B (cm)	H (cm)	V (cm³)	No.	L (cm)	B (cm)	H (cm)	V (cm³)
1				24	11		3		
2		1			12				64
3		2			13		5⅓		
4				72	14	8¼			
5		4			15		7½		
6			5		16				260
7	9				17			4⅓	
8			4		18				58·52
9	10				19	9·5			
10		1			20		4·5		

P. 98   EX. 55

1. x^2	2. $2x^2$	3. $2x^2$	4. $4x^2$	5. $3x^2$
6. $3x^2$	7. $6x^2$	8. $15x^2$	9. $18x^2$	10. $15x^2$
11. $18x^2$	12. $4x^2$	13. $4x^2$	14, $8x^2$	15. $8x^2$
16. $12x^2$	17. $5x^2$	18. $5x^2$	19. $10x^2$	20. $10x^2$
21. $20x^2$	22. $24x^2$	23. $12x^2$	24. $18x^2$	25. $30x^2$
26. $25x^2$	27. $36x^2$	28. $49x^2$	29. $35x^2$	30. $42x^2$
31. $24x^2$	32. $32x^2$	33. $40x^2$	34. $56x^2$	35. $45x^2$
36. $36x^2$	37. $63x^2$	38. $48x^2$	39. $72x^2$	40. $81x^2$

P. 100   EX. 56

1. x^3	2. $2x^3$	3. $2x^3$	4. $2x^3$	5. $4x^3$
6. $4x^3$	7. $4x^3$	8. $6x^3$	9. $6x^3$	10. $8x^3$
11. $8x^3$	12. $24x^3$	13. $24x^3$	14. $12x^3$	15. $48x^3$
16. $20x^3$	17. $50x^3$	18. $60x^3$	19. $48x^3$	20. $30x^3$
21. $40x^3$	22. $90x^3$	23. $60x^3$	24. $15x^3$	25. $72x^3$
26. $10x^3$	27. $120x^3$	28. $24x^3$	29. $60x^3$	30. $36x^3$

P. 101   EX. 57

1. a^2	2. b^2	3. c^2	4. d^2	5. a^3
6. a^3	7. b^3	8. b^3	9. x^2	10. x^3
11. x^3	12. x^4	13. a^4	14. a^4	15. x^4
16. x^4	17. a^5	18. a^5	19. x^5	20. x^5
21. b^6	22. y^6	23. e^6	24. m^6	25. x^5
26. x^6	27. a^5	28. a^6	29. f^7	30. f^7
31. f^7	32. f^7	33. a^2b^2	34. f^2g^2	35. x^2y^2
36. l^2m^2	37. y^2	38. x^6	39. c^7	40. a^7
41. b^2c^3	42. x^8	43. a^6	44. b^6	45. a^6
46. x^6	47. c^{12}	48. $a^2b^2c^2$		

P. 102   EX. 58

1. $4x^2$	2. $9x^2$	3. $16x^2$	4. $4x^3$	5. $9x^3$
6. $16x^3$	7. $6x^3$	8. $6x^3$	9. $12x^3$	10. $8x^3$
11. $8x^3$	12. $12x^3$	13. $6x^4$	14. $9x^4$	15. $12x^4$
16. $8x^4$	17. $10x^4$	18. $15x^4$	19. $20x^5$	20. $10x^6$
21. $12x^7$	22. $6x^6$	23. $12x^5$	24. $16x^6$	25. $24x^6$
26. $24x^6$	27. $24x^6$	28. $24x^7$	29. $60x^9$	30. $40x^8$

P. 103 EX. 59

1. a	2. b	3. x	4. y	5. a
6. b	7. x	8. y	9. a^2	10. b^2
11. x^2	12. y^2	13. a	14. b	15. x
16. y	17. x^3	18. c^2	19. b^3	20. m
21. d^4	22. d	23. $\dfrac{a^2}{b^2}$	24. $\dfrac{b^2}{a^2}$	25. p^4
26. c^3	27. x^3	28. x^3	29. $\dfrac{x^3}{y^2}$	30. e^4
31. k	32. g	33. k^3	34. x^5	35. $\dfrac{a^3}{x^2}$
36. a^5				

P. 104 EX. 60

1. $4x$	2. $3x$	3. $4x^2$	4. $2x^2$	5. $4x^2$
6. $6x^2$	7. $2x^2$	8. $3x^2$	9. x^2	10. $4x$
11. $4x^3$	12. $2x$	13. $2x^3$	14. $6x$	15. $6x$
16. $2x^4$	17. $2x^3$	18. $2x^2$	19. $2x$	20. $3x$
21. $3x^2$	22. $3x^3$	23. $3x^4$	24. $3x^5$	25. $2x^2$
26. $4x^2$	27. $8x^2$	28. $4x^4$	29. $2x^3$	30. $8x^4$

P. 106 EX. 61

1 Interior angles add up to 360°.
2 Any values are possible.
3 The side bisectors and diagonals intersect (cross) at the same point.
4 The diagonals bisect the angles.
5 No. The diagonals do not bisect the angles.
6 No. The diagonals do not bisect the angles.
7 AO = CO = BO = DO. $\angle$AOB = $\angle$BOC = $\angle$COD = $\angle$DOA = 90°. The diagonals are equal and bisect each other at right angles.
8 PX = RX. QX = SX. The diagonals are not equal but they do bisect each other.
9 WA = YA = XA = ZA. The diagonals are equal and bisect each other.
10 (a) In a square the diagonals intersect at right angles.
   (b) In a parallelogram the diagonals do not intersect at right angles.
   (c) In a rectangle the diagonals do not intersect at right angles.

P. 107 EX. 62

1 Each angle = 60°. The angles add up to 180°.
2 $\angle$B = 70°. The sum of the angles is 180°. AC = BC. Triangle ABC is isosceles.
3 The triangle is right-angled.
4 The triangle is acute-angled and also scalene.

5 It is not possible to draw this triangle. The two shorter sides must add up to something greater than the longest side.

6 It is not possible to draw either of these triangles. The sum of the angles should be 180°. (A) The sum is too small — 150° and (B) the sum is too large — 210°.

7 The perpendicular bisectors of the sides meet at the same point. These bisectors also bisect the angles of the equilateral triangle.

8 The perpendicular bisectors of the sides meet at the same point. Only one of them bisects an angle — the bisector of the 5 cm side.

9 The perpendicular bisectors of the sides meet at the same point. **None** of them also bisect an angle.

10 The angle bisectors meet at the same point. Only one of them also bisects a side — the bisector of the angle opposite the 6 cm side.

11 The angle bisectors meet at the same point. None of these bisectors also bisects a side.

12 The angle bisectors meet at the same point. None of these bisectors also bisects a side.

P. 114   EX. 65

1.	324 mg	2.	534 mg	3.	368 mg
4.	7230 mg	5.	4207 mg	6.	6085 mg
7.	30204 mg	8.	5070200 mg	9.	2006005 mg
10.	532 g	11.	864 g	12.	937 g
13.	7203 g	14.	2304 g	15.	2650 g
16.	3·65 g	17.	8·407 g	18.	5·069 g
19.	90·703 g	20.	706·05 g	21.	3004·008 g
22.	4·35 kg	23.	2·064 kg	24.	6·508 kg
25.	2·0403 kg	26.	6·0042 kg	27.	5·008006 kg
28.	0·576 kg	29.	0·070503 kg	30.	0·004069 kg

P. 115   EX. 66

1.	10·94 kg	2.	12·22 kg	3.	10·41 kg
4.	8·59 g	5.	13·9 g	6.	12·07 g
7.	10·74 cg	8.	18·95 cg	9.	10·44 cg
10.	13·138 g	11.	14·813 g	12.	10·134 kg
13.	16·109 g	14.	3·2034 kg	15.	27·128 165 kg

P. 116   EX. 67

1.	343 dag	2.	199 dag	3.	298 dag
4.	289 cg	5.	192 cg	6.	257 cg
7.	79 mg	8.	349 mg	9.	7 mg

10.	67 cg	11.	7601 mg	12.	556 281 cg
13.	5502 cg	14.	27 933·04 g	15.	36 335 mg

P. 117   EX. 68

1.	27·3 g	2.	29 kg	3.	53·6 mg
4.	34 cg	5.	28·8 g	6.	51 kg
7.	33·84 mg	8.	24·15 cg	9.	34·02 g
10.	2·3 cg	11.	2·1 mg	12.	1·8 kg
13.	1·3 kg	14.	1·8 g	15.	2·7 cg
16.	2·2 mg	17.	1·2 kg	18.	2·1 g
19.	51·24 kg	20.	85·6 g	21.	124·08 cg
22.	116·4 mg	23.	78·84 kg	24.	96·5 g
25.	21·92 cg	26.	33·948 mg	27.	169·28 kg
28.	3·2 kg	29.	2·1 cg	30.	5·2 g
31.	9·1 mg	32.	6·4 cg	33.	4·6 kg
34.	4020·5 mg	35.	750 cg	36.	0·22048 kg

P. 118   EX. 69

1.	435 ml	2.	645 ml	3.	479 ml
4.	8340 ml	5.	5308 ml	6.	7096 ml
7.	40305 ml	8.	6080300 ml	9.	3007006 ml
10.	643 l	11.	975 l	12.	848 l
13.	8304 l	14.	3045 l	15.	2760 l
16.	4·76 l	17.	9·508 l	18.	6·078 l
19.	80·804 l	20.	807·06 l	21.	4005·009 l
22.	5·46 kl	23.	3·075 kl	24.	7·609 kl
25.	3·0504 kl	26.	6·0053 kl	27.	6·009007 kl
28.	0·687 kl	29.	0·080604 kl	30.	0·005078 kl

P. 119   EX. 70

1.	14·47 kl	2.	14·34 kl	3.	13·36 kl
4.	10·94 l	5.	17·22 l	6.	15·46 l
7.	13·11 cl	8.	22·08 cl	9.	12·49 cl
10.	15·36 l	11.	17·033 l	12.	12·353 kl
13.	29·219 l	14.	37·5355 kl	15.	40·349066 kl

P. 119   EX. 71

1.	343 dal	2.	199 dal	3.	298 dal
4.	299 cl	5.	199 cl	6.	258 cl

7.	79 ml	**8.**	367 ml	**9.**	7 ml
10.	147 cl	**11.**	8621 ml	**12.**	645171 ml
13.	7411 cl	**14.**	36820·13 l	**15.**	39324 ml

P. 120   EX. 72

1.	27·6 l	**2.**	27·6 kl	**3.**	46·4 ml
4.	48 cl	**5.**	26·1 l	**6.**	57 kl
7.	42·72 ml	**8.**	31·92 cl	**9.**	40·68 l
10.	2·4 cl	**11.**	3·2 ml	**12.**	2·7 kl
13.	3·6 kl	**14.**	4·3 l	**15.**	5·1 cl
16.	6·2 ml	**17.**	2·2 kl	**18.**	4·6 l
19.	64·56 kl	**20.**	96·9 l	**21.**	222·75 cl
22.	208·6 ml	**23.**	98·82 kl	**24.**	178·92 l
25.	34·228 cl	**26.**	38·34 ml	**27.**	287·98 kl
28.	4·3 kl	**29.**	4·6 cl	**30.**	5·3 l
31.	6·1 ml	**32.**	6·4 cl	**33.**	8·2 kl
34.	437·84 ml	**35.**	1100 cl	**36.**	0·35372 kl

P. 122   EX. 73

Question **2**

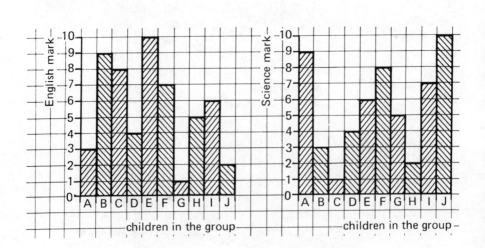

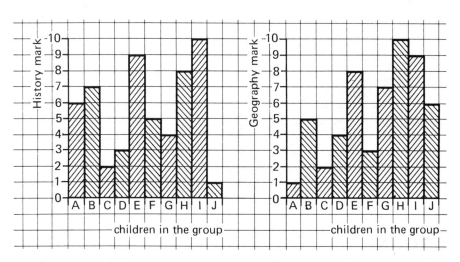

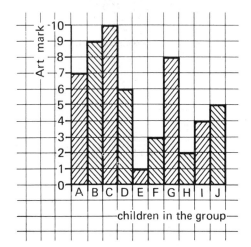

Pupil's name	Total score out of 60
Anthony	34
Betty	36
Charles	30
Dorothy	30
Eric	36
Fiona	36
Gary	30
Harriet	28
Ian	40
Janet	30

Questions
1   Ian gained the highest score.
2   Harriet gained the lowest score.

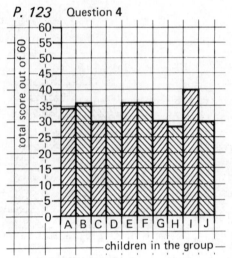

children in the group—

P. *124*   EX. *74*

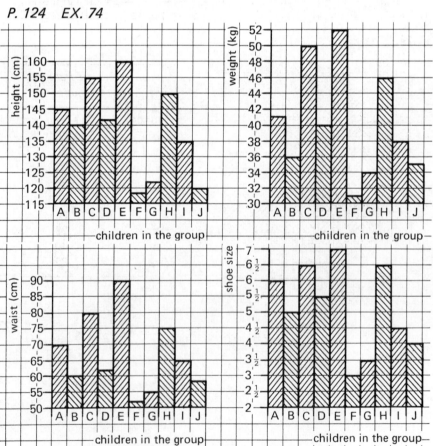

Questions **1** and **2**

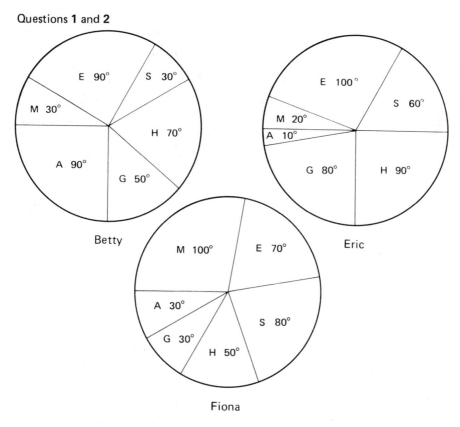

E 90°   S 30°

M 30°

H 70°

A 90°

G 50°

Betty

E 100°

S 60°

M 20°

A 10°

G 80°   H 90°

Eric

M 100°   E 70°

A 30°

S 80°

G 30°

H 50°

Fiona

Questions **3** and **4**

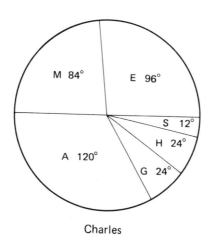

M 84°   E 96°

S 12°

H 24°

A 120°

G 24°

Charles

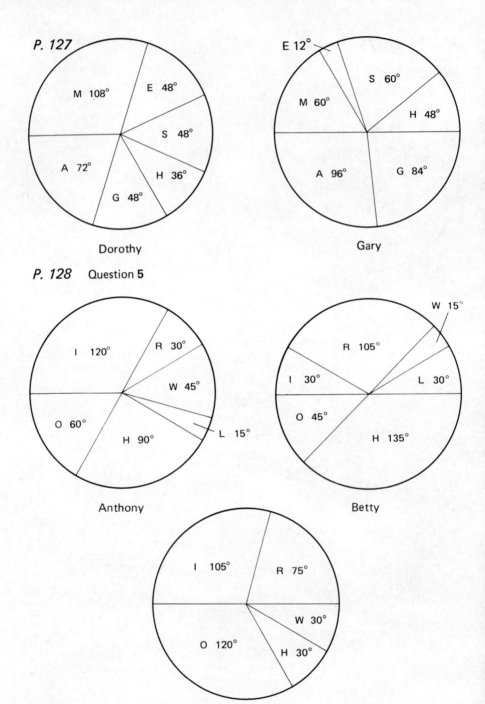

Dorothy

Gary

Question 5

M 108°   E 48°
A 72°   S 48°
G 48°   H 36°

E 12°   S 60°
M 60°   H 48°
A 96°   G 84°

I 120°   R 30°
W 45°
O 60°   H 90°   L 15°

Anthony

W 15″
R 105°
I 30°   L 30°
O 45°
H 135°

Betty

I 105°   R 75°
W 30°
O 120°   H 30°

Charles

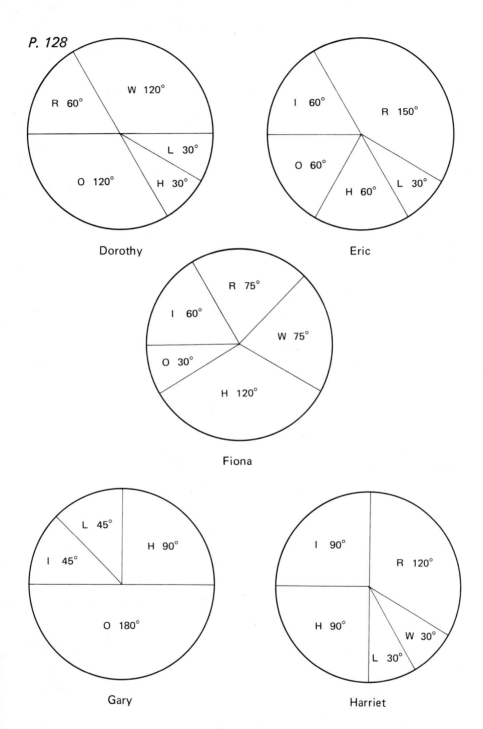

Dorothy

Eric

Fiona

Gary

Harriet

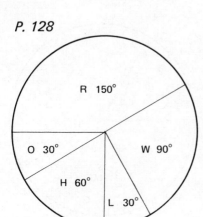

Ian

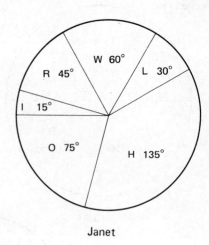

Janet

P. 129   EX. 76

1. 5 cars	**2.** (x + y) cars	**3.** (p + q) dolls
4. 3 books	**5.** (x − y) books	**6.** (p − r) records
7. 1 mile	**8.** (s − v) miles	**9.** (d − f) miles
10. (3 + y) cars	**11.** (4 + q) dolls	**12.** (5 − b) books
13. (35 − r) records	**14.** (5 − v) miles	**15.** (8 − f) miles

16. (a + b − 10) marbles

17. (47 + t + v) autographs

18. Stephen 6 aircraft. Total 9.

19. Stephen 2x aircraft. Total 3x.

20. Christopher (y + 2) engines. Total (2y + 2).

21. Zandra 2s sisters. Total (3s + 2). (Don't forget Carol and Zandra.)

22. Total 11 children. (Don't forget Jeremy and Dominic.)

23. (3x + 3y + 2) children.

24. Marion had 3 sisters.